# Neptune

by Laura Hamilton Waxman

Lerner Publications Company • Minneapolis

Lerner Publications Company
A division of Lerner Publishing Group
241 First Avenue North
Minneapolis, MN 55401 U.S.A.

Website address: www.lernerbooks.com

Words in **bold type** are explained in a glossary on page 30.

Library of Congress Cataloging-in-Publication Data

Waxman, Laura Hamilton.
    Neptune / by Laura Hamilton Waxman.
        p.    cm. – (Our universe)
    Includes index.
    Summary: An introduction to Neptune, describing its
place in the solar system, its physical characteristics, its
movement in space, and other facts about this outer planet.
        ISBN: 0–8225–4655–8 (lib. bdg. : alk. paper)
        1. Neptune (Planet)–Juvenile literature. [1. Neptune
(Planet)] I. Title. II. Series.
QB691.W39 2003
523.48'1–dc21                                    2002000429

Manufactured in the United States of America
1  2  3  4  5  6  –  JR  –  08  07  06  05  04  03

The photographs in this book are reproduced with permission from: NASA, pp. 3, 5, 7, 11, 13, 14, 16, 17, 19, 21, 22, 23, 27; © Hulton-Deutsch Collection/CORBIS, p. 25; © CORBIS, p. 26.

Cover: NASA.

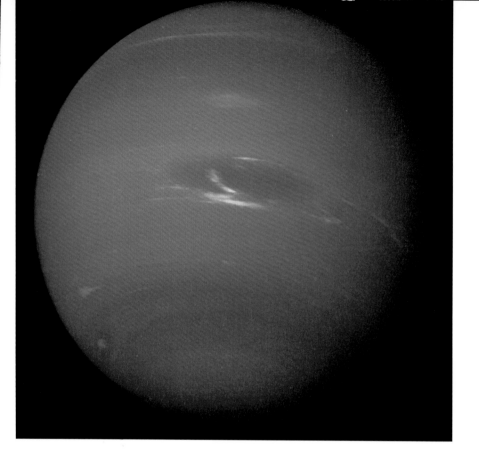

This blue planet has strong winds and storms. It is circled by rings and moons. Do you know which planet it is?

It is Neptune. Neptune is millions of miles away from our home planet, Earth. Neptune is much bigger than Earth. About 60 Earths could fit inside of Neptune.

**Earth**

**Neptune**

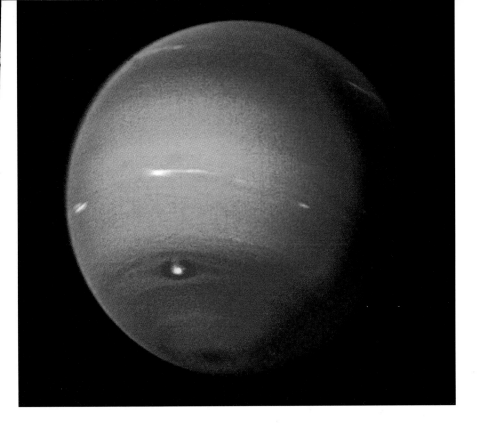

Neptune is a cold and dark planet. It is billions of miles away from the Sun. So it does not receive much of the Sun's warmth or light.

Neptune is part of a family of nine planets called the **solar system.** All of the planets orbit the Sun. To orbit the Sun is to travel around it.

Mercury, Venus, Earth, and Mars orbit closest to the Sun. Jupiter, Saturn, Uranus, Neptune, and Pluto orbit farther out from the Sun.

Neptune is the eighth planet from the Sun. It usually orbits in a path between Uranus and Pluto.

## THE SOLAR SYSTEM

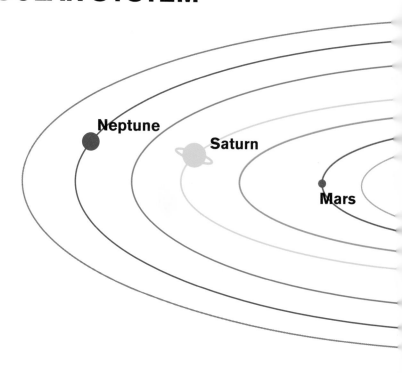

Neptune orbits on an **elliptical** path.
An elliptical path is shaped like an oval.
Most of the planets in the solar system
follow an elliptical path.

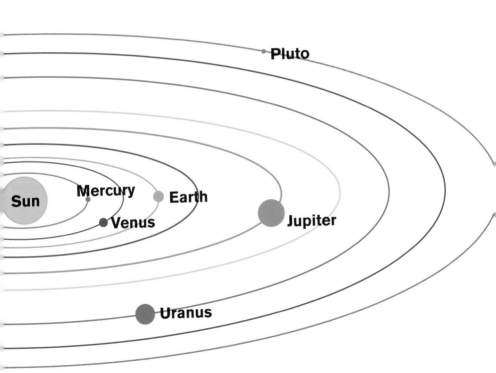

Neptune is much farther from the Sun than Earth is. That makes its path much longer than Earth's path. It takes Neptune about 165 years to orbit the Sun once. Does Neptune move in another way?

Neptune also spins around in space like a top. We call this spinning **rotating.** The time a planet takes to rotate all the way around once is called one day.

Neptune rotates quickly. One day on the planet lasts about 16 hours. One day on Earth lasts 24 hours.

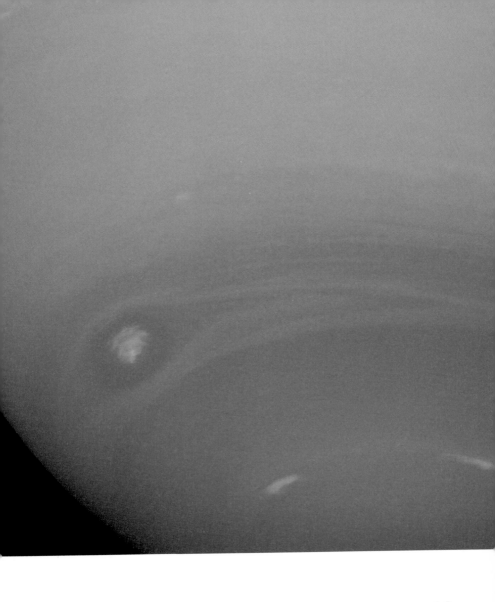

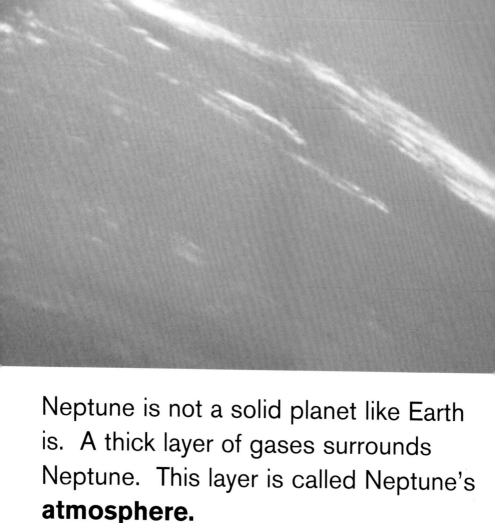

Neptune is not a solid planet like Earth is. A thick layer of gases surrounds Neptune. This layer is called Neptune's **atmosphere.**

Below the atmosphere is a deep ocean.
Neptune's center is probably made of
ice and liquid rock.

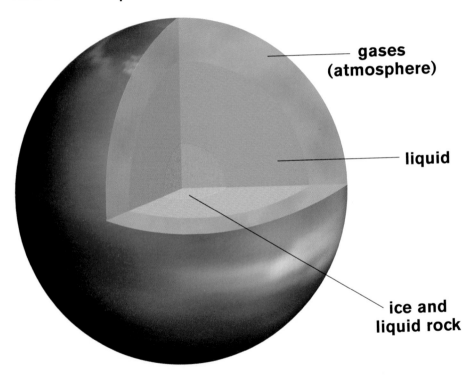

gases
(atmosphere)

liquid

ice and
liquid rock

**NEPTUNE'S LAYERS**

Neptune's atmosphere is a windy and stormy place. Fast winds whip around the planet. The winds carry white clouds with them.

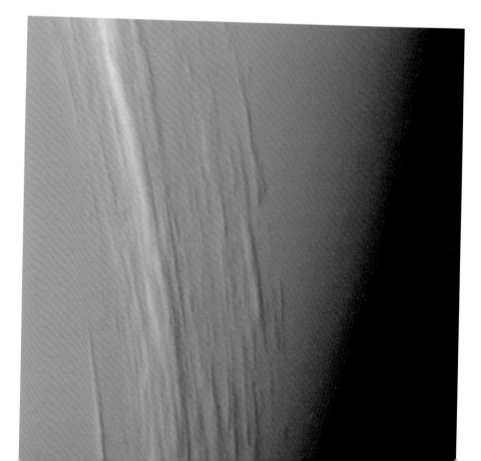

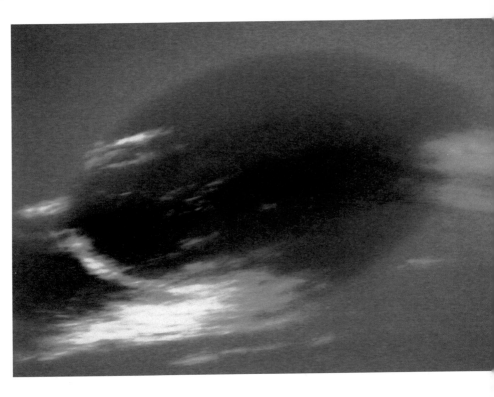

Big storms form in Neptune's atmosphere. This storm was called the Great Dark Spot. It was nearly as big as Earth!

Neptune has four rings that circle it. The rings are probably made of small bits of rock and dust.

Two rings are narrow. They are bright enough to be seen clearly. The other two rings that orbit Neptune are wide and dark. Do you know what else orbits Neptune?

At least eight moons orbit Neptune.
The largest moon is Triton. It is close to
the size of Earth's moon.

The other seven moons are small.
They are Naiad, Thalassa, Despina,
Galatea, Larissa, Proteus, and Nereid.

**Neptune**

**Triton**

The moon Triton is an icy and cold world. It is the coldest place ever found in our solar system.

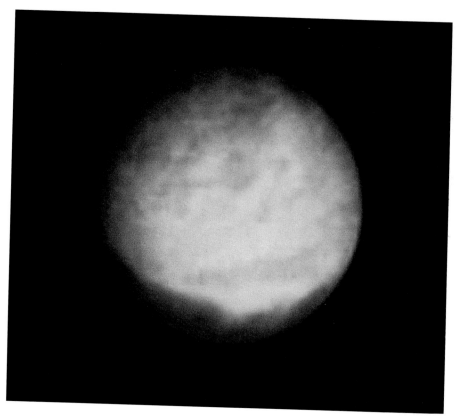

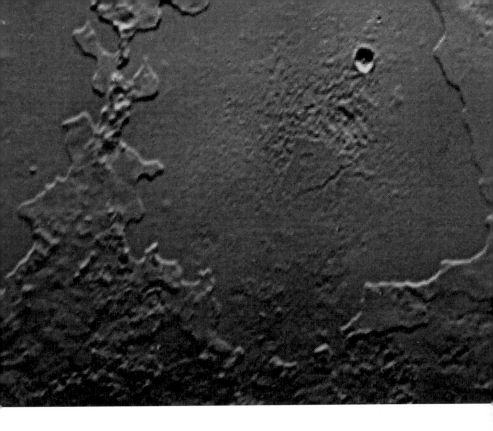

Deep cracks and pits cover Triton. The moon also has frozen lakes and a polar cap. The polar cap is like a frozen mountain.

No one knew about Neptune for thousands of years. Three men worked to discover the planet in 1846. The men were John Couch Adams, Urbain Leverrier, and Johann Galle.

How did people learn more about Neptune?

John Couch Adams

An artist made this picture of *Voyager 2* visiting Neptune

In 1989, a spacecraft from Earth was sent to explore Neptune. The spacecraft was called *Voyager 2.* It took many photographs of the blue planet.

*Voyager 2* taught us about Neptune's atmosphere, rings, and moons. But there is still more to learn about this distant world. What would you like to know?

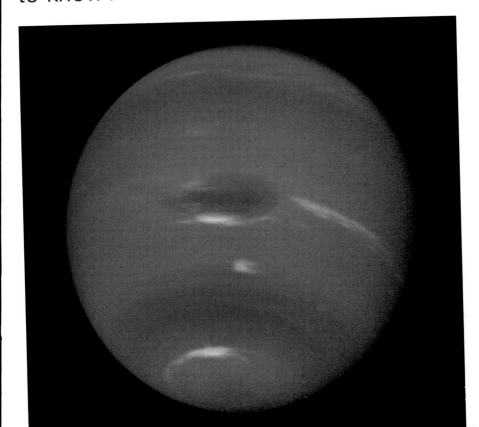

# Facts about Neptune

- Neptune is 2,800,000,000 miles (4,500,000,000 km) from the Sun.

- Neptune's diameter (distance across) is 30,800 miles (49,500 km).

- Neptune orbits the Sun in 165 years.

- Neptune rotates in 16 hours.

- The average temperature on Neptune is −373°F (−225°C).

- Neptune's atmosphere is made of hydrogen, helium, and methane.

- Neptune has 8 moons.

- Neptune has 4 rings.

- Neptune was discovered in 1846 by Johann Galle, with the help of Urbain Leverrier and John Couch Adams.

- Neptune was named after the ancient Roman god of the sea.

- Neptune was visited by *Voyager 2* in 1989.

- At times Neptune orbits farther from the Sun than Pluto does.

- Neptune's moon Triton will someday crash into the planet's atmosphere and be destroyed.

- Winds on Neptune can reach up to 1,500 miles (2,400 kilometers) per hour. These are the fastest measured winds on any planet.

- Neptune is one of four planets called gas giants. The others are Jupiter, Saturn, and Uranus.

# Glossary

**atmosphere:** the layer of gases that surrounds a planet or moon

**elliptical:** shaped like an oval

**orbit:** to travel around a larger body in space

**rotating:** spinning around in space

**solar system:** the Sun and the planets, moons, and other objects that travel around it

# Learn More about Neptune

**Books**

Brimmer, Larry Dane. *Neptune.* New York: Children's Press, 1999.

Simon, Seymour. *Neptune.* New York: Morrow, 1991.

**Websites**

Solar System Exploration: Neptune
*<http://solarsystem.nasa.gov/features/planets/neptune/neptune.html>*
Detailed information from the National Aeronautics and Space
Administration (NASA) about Neptune, with good links to other
helpful websites.

The Space Place
*<http://spaceplace.jpl.nasa.gov>*
An astronomy website for kids developed by NASA's Jet
Propulsion Laboratory.

StarChild
*<http://starchild.gsfc.nasa.gov/docs/StarChild/StarChild.html>*
An online learning center for young astronomers, sponsored by
NASA.

# Index